The Science of Underwater Life

LIVING SCIENCE

Patricia Miller-Schroeder

Gareth Stevens Publishing
A WORLD ALMANAC EDUCATION GROUP COMPANY

For a free color catalog describing Gareth Stevens' list of high-quality books and multimedia programs, call 1-800-542-2595 (USA) or 1-800-461-9120 (Canada). Gareth Stevens Publishing's Fax: (414) 225-0377.

Library of Congress Cataloging-in-Publication Data available upon request from publisher. Fax (414) 225-0377 for the attention of the Publishing Records Department.

ISBN 0-8368-2683-3 (lib. bdg.)

This edition first published in 2000 by
Gareth Stevens Publishing
A World Almanac Education Group Company
1555 North RiverCenter Drive, Suite 201
Milwaukee, WI 53212 USA

Project Co-ordinator: Rennay Craats
Series Editor: Celeste Peters
Copy Editor: Heather Kissock
Design: Warren Clark
Cover Design: Lucinda Cage and Terry Paulhus
Layout: Lucinda Cage
Gareth Stevens Editor: Patricia Lantier-Sampon

Every reasonable effort has been made to trace ownership and to obtain permission to reprint copyright material. The publishers would be pleased to have any errors or omissions brought to their attention so that they may be corrected in subsequent printings.

Photograph Credits:
Bob Allen: page 27 top. Corbis: cover (background). Corel Corporation: cover (center); pages 4 top, 4 bottom, 6 middle left, 6 middle right, 6 far right, 7 far left, 7 middle left, 7 middle right, 7 far right, 8 bottom right, 9 top, 12, 14 top, 14 bottom, 17 bottom left, 17 bottom right, 18 top, 18 bottom, 20 bottom, 23, 25 bottom, 27 bottom. Digital Stock: pages 13 top right, 13 middle left, 15 top right, 16, 19 left, 19 right, 21 bottom, 25 top, 26, 30, 31. Digital Vision: pages 6 far left, 28 top, 28 bottom, 29. Phil Edgell: page 15 top left. Mike Johnson: page 5 bottom. Michael McPhee: page 24 bottom. Tom Stack & Associates: pages 5 top (Thomas Kitchin), 8 top left (John Shaw), 8 bottom left (John Shaw), 9 bottom (Michael S. Nolan), 10 (David & Tess Young), 11 top (Doug Sokell), 11 bottom (Tess Young), 13 top left (Michael S. Nolan), 13 bottom left (David & Tess Young), 13 bottom right (Thomas Kitchin), 20 top (Thomas Kitchin), 21 top (Mike Severns), 22 left (David Young). Visuals Unlimited: pages 8 top right (Gary Meszaros), 13 middle right (Robert F. Myers), 15 bottom (D. Foster), 17 top right (Gary Meszaros), 22 top right (E. C. Williams), 22 bottom right (A. J. Copley), 24 top (Mark E. Gibson).

Printed in Canada

1 2 3 4 5 6 7 8 9 04 03 02 01 00

Contents

What Do You Know about Underwater Life?

Have you ever watched fish in an aquarium? Have you ever walked on a beach and looked at shells or seaweed? If so, you were looking at creatures that make their homes under water.

Coral spends most of its life in one spot.

Frogs spend part of their lives in water and part on land.

Underwater life includes many types of plants and animals. Some, such as goldfish or seals, may be familiar to you. Others may seem very strange. For example, **sea anemones** and sea cucumbers are animals that do not have heads or legs. They look more like flowers than animals.

Salmon swim thousands of miles in their lifetime.

Blue whales are larger than most ships on the ocean.

Activity

Make a List

Make a list of all the underwater life you can name. Compare your list with a friend's list.

Types of Underwater Animals

M any types of animals live under water. Some live in rivers, lakes, or ponds. Others live in oceans.

Types of Underwater Life

Sea Mammals	Fish	Amphibians	Water Insects
• have flippers instead of legs • keep warm with fat called blubber • surface to breathe air	• breathe with **gills** • have scales on body • swim with fins and tail	• adults breathe through moist skin • lay eggs in water • live on land and in water	• have six legs • swim in water or fly above it • young often develop in water

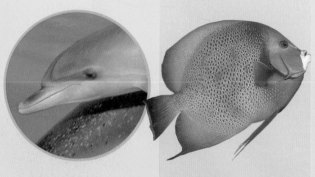  

Examples

dolphins, dugongs, seals, whales	cod, salmon, sharks, trout	frogs, newts, salamanders	dragonflies, fish flies, mosquitoes, water boatmen

Activity

Paint a Picture

Make a painting of underwater animals you would find in either an ocean or a lake. Include at least three different types of animals.

Spiny-skinned Animals	Mollusks	Crustaceans	Coral and Relatives
• may have arms or **tentacles** • have sharp and pointy skins • tube feet cling to rocks	• have soft bodies • most have shells • some have arms or tentacles	• have jointed bodies • have grasping claws or pincers • most have hard shells	• attached to one spot at least part of life • have stinging tentacles • may look like plants

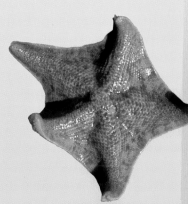   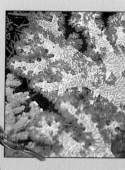

sea urchins, sea cucumbers, starfish	clams, octopuses, sea slugs	crabs, lobsters, shrimps	coral, jellyfish, sea anemones

Life Cycles

All living things have a life cycle. They begin life, grow, reproduce, and die. Some underwater animals, including frogs, change shape as they grow. This type of change is called **metamorphosis**.

A tadpole hatches from an egg laid in water. It breathes with gills. As the tadpole grows, it develops legs and lungs. Soon it becomes an adult frog and lives on land. The frog returns to the water to lay its eggs.

Not all underwater animals go through metamorphosis. For example, **amphibians**, such as frogs, change shape, but mammals do not. Underwater mammals look like their mothers when they are born.

A whale is a mammal. Its young are born under water, tail first. Young whales drink their mother's milk, which contains a lot of fat. This helps them grow quickly. They do not change shape as amphibians do. Whales just get bigger.

Some sea anemones reproduce by splitting in half.

Humpback whale calves are 10–15 feet long (3–4.5 meters) at birth and weigh up to 1 ton (907 kilograms).

Puzzler

Why does a young whale stay with its mother for a long time?

Answer: A young whale must learn many things from its mother. It must learn to communicate with other whales and to travel great distances across the ocean.

Underwater Plants

Many kinds of plants live under water. Some are so tiny you cannot see them without a **microscope**. These plants are called **phytoplankton**. They grow near the surface of the water where sunlight reaches.

Sea grass grows in shallow water near coastlines. It has flat, green blades like land grass. Sea grass covers large areas. It makes underwater pastures where other plants and animals live and feed.

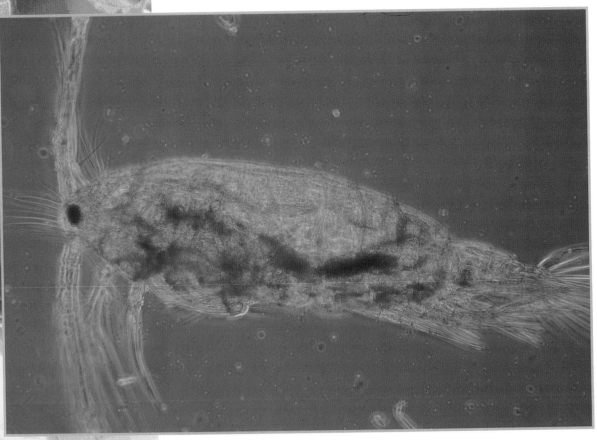

Phytoplankton provide food for swarms of tiny animals, including this **zooplankton**.

Algae are also common water plants. Most algae are tiny. They can be green, blue-green, or red. Large algae are called seaweeds. They are red, green, or brown and grow in long strands. Some anchor themselves to rocks. Others float freely in the water. Seaweeds provide food and homes for many underwater animals.

Sometimes algae forms floating mats, or blooms.

Sea otters wrap themselves in kelp when they sleep. This heavy seaweed keeps them from drifting out to sea.

Puzzler

Sometimes very large blooms of algae form. What problems does this cause for a pond or marsh?

Answer: The bloom shades underwater plants, so they begin to die. This destroys the food and shelter of underwater animals. The bloom also takes oxygen from the water, so fish cannot breathe.

Watery Webs

Most underwater plants and animals eat other living things to survive. Sunlight or dissolved nutrients feed other organisms. They are all part of a food chain. Energy is passed along when one thing eats another. Connected food chains make a food web.

Animals and plants, including this frog, provide energy for other animals.

Zooplankton are eaten by small fish. Small fish are eaten by larger fish. Larger fish are eaten by sea mammals, such as dolphins. In turn, sea mammals and large fish are eaten by killer whales. Zooplankton are also eaten by sharks and some large whales, including humpback whales.

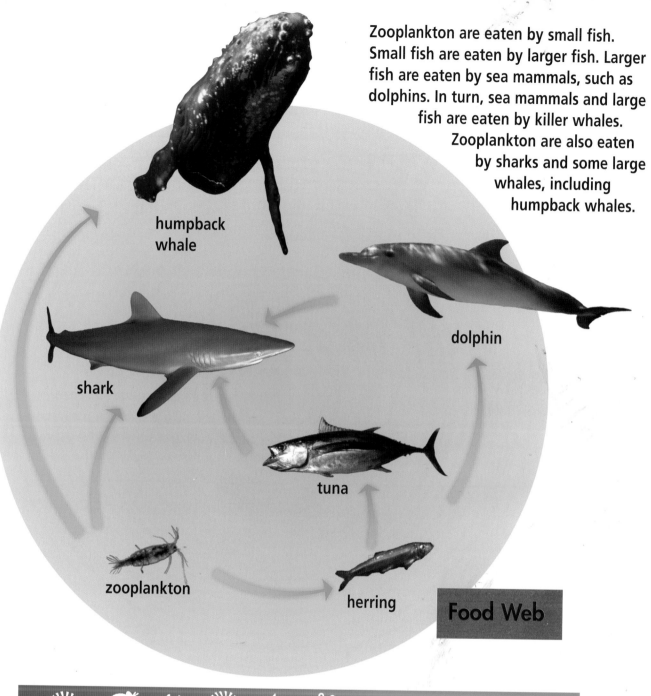

humpback whale

dolphin

shark

tuna

zooplankton

herring

Food Web

Puzzler

Does anything eat killer whales?

Answer:
Yes. When killer whales die, their bodies decay. This provides food for phytoplankton, zooplankton, and scavengers, including crabs.

At Home Under the Water

Water provides many different **habitats**. Some are in the fresh water of rivers, lakes, and wetlands. Others are in the salty water of oceans.

Wetlands are places where water and land meet. They include ponds, marshes, and swamps. These places provide food and shelter for a wide variety of creatures.

Wetlands are home to frogs, salamanders, fish, turtles, snakes, and insects. Ducks, geese, herons, loons, and other birds feed and hunt in the water.

Many plants and animals live in the sunny upper ocean. Coastal regions have huge areas of waving sea grass. Many small fish and other creatures, such as green sea turtles, dugongs, and waterfowl, live here and feed on the grasses.

The deepest ocean is dark and cold. Some animals living here have no eyes. Others have mouths bigger than their bodies. On the ocean floor, cracks in Earth's crust spurt out very hot water that contains minerals. This allows bacteria to grow. Tube worms and clams move in to eat the bacteria. Other creatures come to eat the worms and clams. Soon a busy **vent community** forms around the cracks.

Puzzler

Why are dugongs sometimes called sea cows?

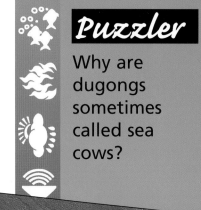

Underwater life, such as the rattail fish, swims slowly at the bottom of the ocean. They hunt for small animals, such as these tube worms, to eat.

Answer: Dugongs are large, fat animals that graze on the sea grass meadows like cows in a land pasture. Actually, they act more like pigs under the water in some ways. They often pull up and eat the tough roots of the grass.

15

Moving and Breathing

Underwater animals have **adapted** to moving and breathing in watery habitats. Many use fins or flippers to move. Fins are usually skinlike material that stretches out like a sail. A fish may have fins on its tail, back, or sides. Flippers are limbs that work like paddles. They are better than legs for swimming.

Sea mammals, such as sea lions, swim using flippers.

An insect called the water boatman carries a layer of air under its wings. This lets it breathe under water like a scuba diver.

Whales and dolphins have a blowhole on top of their heads. This allows them to breathe at the water's surface. They spout out used air and bring in fresh air. Gills are special organs that let fish and many other animals breathe under water. These animals draw oxygen from the water, so they do not have to go to the surface.

Activity

Listening to Whale Sounds

Listen to recordings of whale sounds. Close your eyes and imagine you are traveling under water. What do you hear and see?

Whales breathe through blowholes connected to their lungs. They cannot breathe through their mouths. Their mouths are connected directly to their stomachs.

Many underwater animals are bullet- or wedge-shaped. This lets them move easily through water.

Seafood Feast

Underwater hunters have many ways to find and trap prey. Some use stinging tentacles. Tentacles are long and stringy. They are used to trap and kill food.

Animals that float in the water, such as jellyfish, have tentacles, too.

Animals that stay in one spot often have tentacles. Corals are good examples.

Some types of whales and dolphins use **echolocation** to find food. They bounce sound waves off objects in the water. The returning echo helps them locate schools of fish or other prey.

Other animals travel to catch their food. Sharks are strong swimmers. They travel long distances when they sense blood or the movements of splashing prey. Sharks can smell one drop of blood in 115 quarts (109 liters) of water.

Sharks are one of the most feared animals in the ocean. People fear sharks, too, but there are fewer than one hundred shark attacks worldwide per year.

Activity

Design a Predator
Design your own underwater predator. Use pieces of colored construction paper and paints or crayons. Show what your predator would eat and how it would catch its prey. Be sure to place it in a habitat and illustrate its prey.

Dolphins play together and hunt for food in groups. These fast swimmers can chase and catch other fish for lunch.

Avoiding Danger

Animals that live under water must constantly avoid danger. Some stay safe by making themselves hard to see using **camouflage**. Others taste bad or use other animals to defend themselves.

Camouflage
helps animals disappear.
Their color, shape, or pattern blends into the background. Beluga whales live in cold, arctic waters. Their white bodies look like floating chunks of ice.

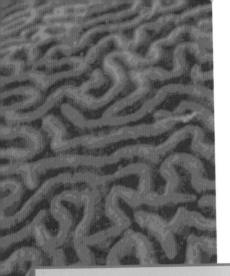

Hiding
keeps some animals safe.
A large octopus can squeeze into a rock crack only 3 inches (7 centimeters) wide. How? Its body has no bones! An octopus can also squirt ink into the water and hide behind the ink cloud.

A Mucous Coat makes some creatures taste very bad.

This slimy covering also protects them against stings or poisons. Clown fish live among the stinging tentacles of sea anemones. These tentacles, along with the fish's coat, keep predators away.

Puzzler

Why are some fish a dark color on their back and a light color on their underside?

Answer:
This is a form of camouflage. Predators from above look down into the dark water. The fish's back blends into the dark. Predators from below look up toward the surface. The fish's belly blends with the light water above it.

Swimming in schools helps many fish avoid enemies.

The fish swim closely together in groups called schools. They all look alike and move at the same time. This makes it hard for a predator to pick out one fish.

Living Fossils

A fish called the coelacanth (SEE-la-kanth) is a living fossil. Scientists thought this fish had been **extinct** for 65 million years. They were wrong. A live one was caught in an African river in 1938. In 1952, another coelacanth was caught in the Indian Ocean. More have been caught since then.

Coelacanths are large, slimy, blue or brown fish. They weigh about 200 pounds (90 kilograms). Nobody knows how many members of this ancient species are alive today.

A coelacanth today does not look very different from the fossil of an ancient coelacanth.

Some animals that lived under water millions of years ago have close relatives alive today. Many modern sharks are very similar to ones that lived 300 million years ago. They have been successful hunters and scavengers for a very long time.

Sea turtles similar to today's hawksbill turtles traveled the oceans 200 million years ago.

Activity

Underwater Life Then and Now

Look at ancient underwater animals in books or a museum. List ways the ancient animals are the same as underwater animals today. List how they are different.

Fishing for Knowledge

People have been fascinated by underwater life for centuries. Some people keep underwater animals as pets. They set up small home aquariums. Scientists set up huge aquariums and marine parks. This lets them study underwater life up close.

Goldfish and small tropical fish make good pets if they are cared for properly.

Some scientists dive deep into oceans and lakes. They learn how fish and whales behave in their natural habitat. Often, scientists record and study the sounds underwater animals use to communicate.

Many people travel the world to dive with and study sharks.

Activity

Study a Fish

Watch fish swimming in an aquarium. Pick one fish and watch it for five minutes. Describe everything about it. What does it look like? How does it behave? Sprinkle a little fish food in the water. What happens? What special features does your fish have? How do these features help it live under water?

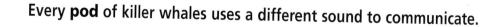

Every **pod** of killer whales uses a different sound to communicate.

Fishy Careers

Y ou may want to learn more about underwater animals and plants by working with them. Marine biologists study life in the oceans. Freshwater biologists study life in ponds, lakes, and rivers. You must go to college to become a biologist. Many biologists work at large aquariums. Animal trainers and divers also work at aquariums.

Some marine biologists dive with sharks. They watch shark behavior and learn more about how sharks live.

Fish farmers raise fish, such as trout or salmon. They keep the fish in protected ponds or underwater corrals.

Activity

Do Your Own Research

Ask a parent or teacher to help you learn about these interesting careers that involve underwater life:

- animal trainer
- aquarium worker
- biologist
- diver
- fish farmer
- underwater photographer

Underwater photographers help others see what underwater life is like.

Threats to Underwater Life

Many of Earth's oceans, lakes, rivers, and ponds are becoming **polluted**. Pollution is dangerous to underwater life. Harmful chemicals poison many plants and animals. People sometimes dump these poisons into water to get rid of them. Chemicals can also leak into water from factories and farmers' fields.

Oil spills from tanker ships pollute water and kill underwater life. Some animals die right away. Others may starve if their food dies. Those that survive often cannot breed or raise healthy young.

Pollution, such as oil spills, hurts animals that live in and near the water. Even with rescue efforts, some animals die.

Plastic bags and broken nets injure and kill many underwater animals. These objects can trap or choke animals. Sea turtles that try to eat jellyfish often eat floating plastic by mistake. This can kill them. Underwater mammals sometimes drown in nets.

Fishermen use nets to catch fish. Sometimes dolphins become tangled in the nets, too.

Puzzler

Can warm water count as pollution?

Answer:
If it is dumped into a cool river or lake, yes. Warm water contains less oxygen than cool water. Cold-water fish find it difficult to breathe in warm water.

Protecting Underwater Life

Many types of underwater plants and animals are **endangered**. Pollution has destroyed much of their habitat. Hunters have also harmed some animal populations. Certain types of large whales are nearly extinct. Hunters have killed too many of them. Hunters also kill seals and sea otters for their skins.

Fishing can also threaten a group's survival. Certain types of fish are nearly extinct due to over-fishing by humans.

Many people try to save underwater life. Certain groups, such as Greenpeace, work to protect ocean life and habitats. Some public aquariums are raising and releasing endangered sea turtles.

Bans against hunting fur seals or selling fur seal products have helped rebuild the population. These laws give this seal pup a good chance of surviving.

You can help save underwater life, too. Write a letter to one of the groups below. Ask what they are doing to help underwater life. Find out if they have activities in which you could get involved.

Greenpeace
1436 U Street NW
Washington DC
20009
USA

www.greenpeaceusa.org

Center for Marine Conservation
1725 DeSales Street NW
Suite 600
Washington, DC
20036
USA

www.cmc-ocean.org

Wild Killer Whale Adoption Project
Vancouver Aquarium
Marine Science Centre
P.O. Box 3232
Vancouver,
British Columbia
V6B 3X8
Canada

www.whalelink.org/main.html

Scuba divers are invited to swim with underwater life. They are also asked not to disturb the areas they visit.

Glossary

adapted: suited to a certain way of life.

amphibians: animals that live in water and on land.

camouflage: color, shape, or pattern that blends into the background.

echolocation: the use of echoes to locate food or objects.

endangered: at risk of dying out completely.

extinct: no longer living.

gills: organs that allow an animal to breathe under water.

habitat: the place where an animal or plant lives.

metamorphosis: a change in appearance or form as an animal grows.

microscope: an instrument that makes tiny objects look larger.

phytoplankton: tiny floating plants that live in water.

pod: name for a family group of whales.

polluted: containing dirt, garbage, or other harmful substances.

sea anemone: an animal with stinging tentacles that lives in the ocean and resembles a flower.

tentacles: long limbs from an animal's body that are used as feelers.

vent community: a group of animals that lives around a crack or vent in the ocean floor where very hot water pours out.

zooplankton: tiny floating animals that live in water.

Index

Web Sites

www.panda.org/kids/wildlife/idxocmn.htm

www.whalelink.org/main.html

www.cccturtle.org

www.frog.simplenet.com/froggy

Some web sites stay current longer than others. For further web sites, use your search engines to locate the following topics: *fish, marine biology, marine life, oceans, sharks,* and *whales.*